AF484680

FACT CHECKERS
PLANT
MYTHS
&
MISCONCEPTIONS
40+ Amazing Facts About Plants
By Carrie Rodell and Kizzi Roberts
LearningSpark EDUCATIONAL PUBLISHING

HAVE YOU EVER WONDERED???
DO ALL PLANTS GROW FROM SEEDS?
DO SUNFLOWERS FOLLOW THE SUN?

# DO HOUSE PLANTS CLEAN THE AIR?

## FACT FILE

This book is full of myths and misconceptions about plants. Lucky for you it's also full of wondrous facts. Fact checkers is about setting the record straight once and for all.

## FACT CHECKERS SAY...

## KEEP READING!

# DO FLOWERS ALWAYS SMELL GOOD?

## WATER-LOGGED

Many cactus plants live in dry, hot climates such as the desert. Most plants cannot survive in a place with little rainfall because they need water. Can a cactus survive without water?

### FACT FILE

Some people say *cacti* is the correct plural noun to use when talking about more than one cactus. Others say *cactuses*. Both terms are correct.

### DO SPIDERS LIVE INSIDE CACTUSES?

You may have heard the story of a cactus that starts shaking and eventually explodes, releasing hundreds of baby tarantulas into a woman's house. Thankfully, this is only a legend. Although a spider might lay its eggs *on* a cactus, no spider is known to lay its eggs *inside* a cactus.

# NOPE!

A cactus can survive for a long time on the water it stores in its stem, but eventually it will die without water. Some cacti can live for more than a year on the hundreds or even thousands of gallons of water they store.

### ARE CACTUSES EDIBLE?

Yes! A few cactus varieties can be eaten, such as the prickly pear in the picture above. People eat both the thick, flat pads and the fruits, after removing the spines and thorns, of course.

NICE TO MEAT YOU
Does the Venus flytrap eat meat?

FACT CHECKERS SAY...

# ACCURATE!

The plant does eat meat. Like animals that are carnivores, it digests animals for nutrients. The Venus flytrap digests insects and even small frogs or snakes once in a while. There are more than 700 carnivorous plant species!

### DOES THE VENUS FLYTRAP EAT MOSTLY FLIES?

Actually, the plant lives on ants and spiders mostly. Insects crawling across the lobe are much more common than flying insects landing on it.

### IS THE VENUS FLYTRAP AN EXOTIC PLANT?

The Venus flytrap may look like an exotic plant that grows in the tropics, but this strange plant is native only to a small area of wetlands and forests in North Carolina and South Carolina.

### WILL THE VENUS FLYTRAP DIE IF IT DOES NOT CATCH INSECTS?

No. The plant gets extra nutrients from insects, but it also gets energy from the same source as most plants: the sun.

The lobes that trap insects also trap sunlight and turn it into food through photosynthesis.

## FACT FILE

Each lobe of the Venus flytrap has three small hairs on it. When an insect touches two of these hairs in a row, two lobes close like jaws and trap the insect between them. The trap may also close if the same hair is touched twice.

SO LONG, SOIL
Do all plants need soil to grow?

FACT FILE
The water lily is a type of aquatic plant that lives in fresh water such as ponds and lakes.

NOPE!
Plants need water, light, and nutrients to grow. While many plants get those nutrients from soil, some plants can absorb nutrients from the air or from water.

GREEN SCHEME
Are all plants green?

NOPE!
This ghost plant appears white because it does not produce chlorphyll, which is what makes plants appear green.

FACT CHECKER REPORT
Plants that do not produce chlorophyll cannot make their own food through photosynthesis, so they must absorb nutrients from nearby organisms.

MOVING PARTS
Can plants move?

YES!
While plants don't walk around like animals, they can move certain parts such as stems or leaves in response to touch, light, or other stimulus.

SUNRISE, SUNSET
Do sunflowers follow the sun?

SOMETIMES.
The heads of young sunflowers change direction throughout the day as they turn to follow the sun from east to west. This is called heliotropism. When the plants are more mature, they stop following the sun and the flower heads face east.

### DO INDOOR PLANTS CLEAN THE AIR?

Yes, but not very well. You would need between 100 and 1,000 plants in your bedroom to clean the air well. That's a lot of plants!

### DO CATS REALLY LOVE THE SMELL OF CATNIP?

Catnip is an herb from the mint family. It is known for having an effect on cats. When they smell catnip, 50 to 70% of adult cats rub their heads on it, make noises, drool, or roll around. Some cats do not react to catnip at all, though. The behavior seems to be inherited (or not).

### DO ALL FLOWERS SMELL GOOD?

Some flowers have no scent at all, and others smell terrible! This stinky corpse flower has an unpleasant odor that smells like rotting meat. The smell attracts pollinators like flies and carrion beetles that love dead and decaying matter.

TOMATO, TOMATO

To-MAY-to, to-MAH-to...not only do people pronounce tomato differently, they also argue about how to classify this tasty fruit...or vegetable. Are tomatoes vegetables?

FACT CHECKERS SAY...

## TECHNICALLY NO, BUT...

If you ask a scientist, they might say a vegetable is a plant part that people eat, while a fruit grows from a flower and has seeds inside. By this definition, tomatoes are fruits, but in 1893, the United States Supreme Court got involved in the debate over whether a tomato is a fruit or a vegetable. The court ruled that although scientists classify it as a fruit, it is a vegetable to the average person.

# FACT CHECKER REPORT

### ARE TOMATOES POISONOUS?

No! A long time ago, some people thought tomatoes were poisonous for two reasons:

1. Tomatoes are members of the nightshade family. This plant family includes many plants that are safe to eat. But it also includes belladonna (also known as deadly nightshade), which *is* a poisonous plant.

2. Some people got sick after eating tomatoes, but it wasn't because the tomato was poisonous. Tomatoes served on pewter plates pulled lead out of the plate. When people ate the tomatoes served on these plates, they would eat a lot of lead too. The people were sick from lead poisoning, not because of the tomatoes.

### ARE ALL TOMATOES RED?

No, tomatoes come in many shapes, sizes, and colors. Gardeners love to grow red, orange, yellow, green, and purple tomatoes. There are even pink tomatoes, black tomatoes, and striped tomatoes!

GOING BANANAS
Do bananas grow on trees?

NOPE!
The banana plant is actually a giant herb! The plant does not have a woody trunk like a tree. Instead, it has a large non-woody stem.

# WANTED WEEDS

Are all weeds useless, ugly plants?

## NO.

A weed is just a plant growing in a place where humans don't want it to grow. Some weeds like dandelions are edible, but you should never eat anything without checking with an adult first.

## FACT FILE

Dandelions are full of good nutrients people need. Dandelions have more vitamin A than spinach, more vitamin C than tomatoes, and are also a good source of iron, calcium, and potassium.

## FACT CHECKER REPORT

### ARE PEANUTS REALLY NUTS?

Peanuts are not nuts. They are legumes, which means they are more closely related to beans than to nuts.

# SUNLIGHT AND SPROUTS

Do seeds need a dark place to sprout?

FACT CHECKERS SAY...

## FACT CHECKER REPORT

### DO PLANTS NEED OXYGEN?

Yes. Like animals, plants need oxygen to turn sugars into energy. Plant roots also take in oxygen from pockets of air in the soil. If the soil becomes too soggy from water, plant roots can suffocate and die.

Plants use carbon dioxide during photosynthesis. This process uses carbon dioxide, water, and sunlight to make sugar and oxygen. So, even though plants use oxygen, they make more oxygen than they use.

## NOT ALWAYS.

Most seeds sprout underground in the dark, but some seeds require sunlight to sprout. One example is a seed from the empress tree (in the above photo). Seeds that require sunlight may be scattered on top of the soil instead of being planted and covered with soil.

JUST A FUNGI
Are wild mushrooms toxic?

FACT CHECKERS SAY...

SOMETIMES!
Some wild mushrooms are edible, but many kinds are toxic. To be safe, people should never eat unfamiliar mushrooms.

# FACT CHECKER REPORT

### CAN MUSHROOMS GROW IN DARK PLACES?

Yes. Since they get their food from other organisms, mushrooms do not need sunlight to make food through photosynthesis. They can thrive in dark areas like caves or forest floors.

### ARE MUSHROOMS AND TOADSTOOLS THE SAME?

A toadstool is a type of mushroom. Many people use *toadstool* to describe a mushroom that is poisonous or can't be eaten, but there is no single mushroom variety named *toadstool*.

### ARE BRIGHTLY COLORED MUSHROOMS POISONOUS?

Not necessarily. Some poisonous mushrooms are indeed colorful, but many deadly mushrooms are plain brown or white.

SO MUCH TO SAY
Do plants grow faster when you talk to them?
FACT CHECKERS SAY...

## NOT REALLY, BUT...

While there is no scientific evidence to support lucky (or unlucky) plants, some cultures believe certain plants attract luck, protection, wealth, and even love. The money tree's braided trunk is believed to trap good luck.

## FACT CHECKER REPORT

### DO PLANTS COMMUNICATE?

Plants don't use words like humans, but they still have ways to send messages. Plants use hormones, electrical signals, and other chemicals to send messages. These silent signals deliver information from one part of the plant to another, and even between plants.

## MAYBE.

Scientists aren't sure yet. There is some evidence that plants may grow better in response to any sound vibration, not just human speech. Other people wonder if it's the carbon dioxide we exhale that plants are responding to.

PLANTING A SEED
Do all plants require seeds to reproduce?
FACT CHECKERS SAY...

# FACT CHECKER REPORT

## DO SEEDS ALWAYS GROW INSIDE FRUIT?

Not always. Strawberry seeds are on the outside of the fruit. Also, pine cones contain seeds, but pine cones are definitely not fruit!

## DOES FRUIT DEVELOP FROM A FLOWER?

Yes, all fruit comes from flowers, but not all flowers make fruit.

## NOT NECESSARILY.

Tulips, daffodils, and many flowers reproduce from bulbs. Others, like some strawberry plants, send out tiny "child" plants along the ground called runners. Even though these plants also produce seeds, the seeds aren't required for the plant to reproduce.

# GLOSSARY

Definitions for selected words found in this book.

**AQUATIC** (adjective) living or growing in water

**CARNIVORE** (noun) any plant or animal that feeds on animals

**CARRION** (noun) dead and rotting flesh

**CHLOROPHYLL** (noun) the green coloring of leaves and plants that is important to photosynthesis

**CLIMATE** (noun) the long-term weather conditions of an area

**COMMUNICATE** (verb) to exchange information with signals

**DECAYING** (adjective) rotting or breaking down

**DIGEST** (verb) to convert food into energy that can be absorbed and used by the body

**EDIBLE** (adjective) something that is safe to eat as food

**HELIOTROPISM** (noun) the movement of plants or plant parts in response to light from the sun

**HERB** (noun) a plant used for its flavor, scent, or for medicinal purposes

**KINGDOM** (noun) in biology, a category of the second highest rank used to organize living things

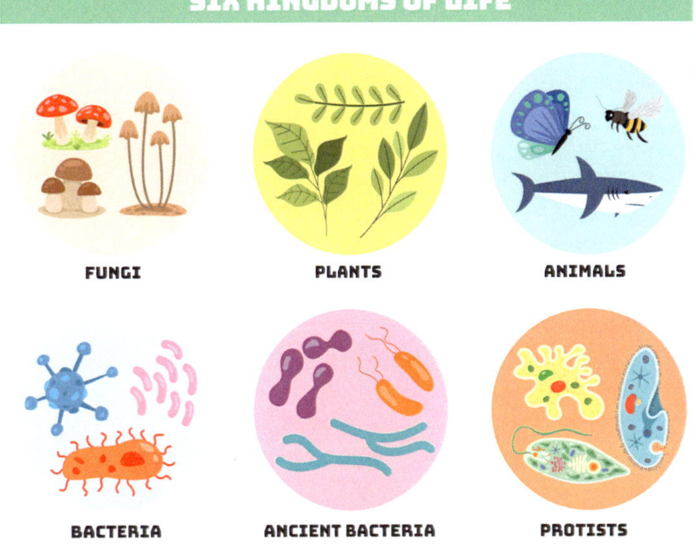



**LEGEND** (noun) a story not documented in history, but it continues to be told and is often accepted as true

**LEGUME** (noun) a plant that has a fruit of a seed pod; examples: beans, peas, and clover

**LOBE** (noun) a roundish part of an organ or a leaf

**MATURE** (adjective) complete in natural growth or development

**NATIVE** (noun) a species originating in a particular place or area

**NUTRIENT** (noun) any substance that nourishes an organism

**ORGANISM** (noun) a form of life made of different parts that work together to maintain vital processes

**PARASITIC** (adjective) living as a parasite on or in a host organism and taking nutrients from the host, causing harm

**PHOTOSYNTHESIS** (noun) the process which allows green plants to use sunlight to create food from carbon dioxide and water

**PLURAL** (adjective) more than one

**RESPIRATION** (noun) the process of taking in oxygen and releasing carbon dioxide in order to produce energy

**STIMULUS** (noun) something that causes an action

**TART** (adjective) having a sharp, sour taste

**THRIVE** (verb) to grow and healthy

**TOXIC** (adjective) having the effect of poison

**WETLAND** (noun) an area of land that is swampy or marshy

Text © 2026 Kizzi Roberts and Carrie Rodell
Book design and layout by Kizzi Roberts
Edited by Carrie Rodell
Published by Learning Spark Educational Publishing in Rogersville, Missouri
Hardcover ISBN: 979-8-88884-037-5
Paperback ISBN: 979-8-88884-038-2
Ebook ISBN: 979-8-88884-044-3
Library of Congress Control Number: 2026905756



Photographs © Hilda Weges/depositphotos.com; norikazu/depositphotos.com; A1804/depositphotos.com; lzf/depositphotos.com; Olgallinich/depositphotos.com; weerapat/depositphotos.com; fisher05/depositphotos.com; thug1747/depositphotos.com; wirestock_creators/depositphotos.com; alessandrozocc/depositphotos.com; Gudella/depositphotos.com; elisalocci/depositphotos.com; gitanna/depositphotos.com; mario_plechaty_photography/depositphotos.com; sanddebeautheil/depositphotos.com; Erida/depositphotos.com; seekeaw@gmail.com/depositphotos.com; panattar/depositphotos.com; umuller/Envato; yanadjana/Envato; mblach/Envato; rthanuthattaphong/Envato; natanavo/Envato; Sepaolina/Envato; mblach/Envato; yusufdemirci/Envato; alessandrozocc/Envato; Sergii Figurnyi/Adobe Stock; helivide/Adobe Stock.

9 7 9 8 8 8 8 8 4 0 3 7 5